Backyard Livestock

A Prepper's Guide to Raising Chickens, Goats, Sheep, and More for Food and Fertilizer

LEIGH BIRD

Table of contents

Table of contents... 3
INTRODUCTION.. 1
    About Backyard Livestock............................ 1
    Benefits of Raising Livestock for Preppers.............. 2
    Understanding the Prepper Mindset....................... 3
CHAPTER 1... 6
Getting Started with Backyard Livestock.................. 6
    Assessing Your Space and Resources.................... 6
    Legal Considerations and Zoning Regulations......... 7
    Budgeting for Initial Costs............................. 8
    Implementation of Spatial Considerations............. 10
    Navigating Legal Considerations....................... 11
    Fine-Tuning the Budgeting Process.................... 12
    Navigating Ahead...................................... 13
CHAPTER 2.. 14
Selecting the Right Livestock for Your Needs......... 14
    Chickens: Breeds, Housing, and Care.................. 14
    Goats: Types, Feeding, and Shelter.................... 16
    Sheep: Choosing the Right Breed, Grazing, and
    Health.. 17
    Navigating Ahead...................................... 18
CHAPTER 3.. 20
Planning and Building Livestock Infrastructure......20
    Coop and Barn Designs................................ 20
    Fencing Options and Considerations.................. 21

Watering Systems and Waste Management...........23
Navigating Ahead... 24
CHAPTER 4.. **25**
Feeding and Nutrition..**25**
Understanding Livestock Nutritional Needs...........25
Creating Balanced Diets...................................... 27
Seasonal Considerations in Feeding..................... 28
Navigating Ahead... 29
CHAPTER 5.. **31**
Health and Veterinary Care................................ **31**
Identifying Common Ailments............................... 31
Vaccination Protocols.. 32
Routine Health Checks... 34
Navigating Ahead... 35
CHAPTER 6.. **37**
Breeding and Reproduction................................ **37**
Understanding Animal Reproductive Cycles.......... 37
Selective Breeding for Traits.................................38
Pregnancy and Birthing.. 40
Navigating Ahead... 41
CHAPTER 7.. **43**
Daily Care and Maintenance...............................**43**
Daily Chores and Routine Tasks............................43
Grooming and Hygiene Practices.......................... 44
Dealing with Emergency Situations....................... 46
Navigating Ahead... 47
CHAPTER 8.. **49**

Maximizing Yield: Eggs, Milk, and Meat................. **49**

    Egg Production Tips and Tricks............................ 49

    Milking Techniques and Dairy Management.......... 51

    Slaughtering and Processing Meat....................... 52

    Navigating Ahead... 53

CHAPTER 9... **55**

Utilizing Livestock Products............................ **55**

    Composting with Animal Waste............................ 55

    Using Manure as Fertilizer.................................. 56

    Other Creative Ways to Utilize Livestock Byproducts. 58

    Navigating Ahead... 59

Chapter 10... **61**

Sustainable Practices and Future Considerations.. **61**

    Implementing Eco-Friendly Practices.................... 61

    Long-Term Planning for Sustainable Livestock....... 63

    Expanding or Diversifying Your Livestock Endeavors. 65

    Navigating Ahead... 67

CONCLUSION.. **69**

    Recap of Essential Points................................... 69

    Encouragement for Ongoing Success in Backyard Livestock Management... 71

INTRODUCTION

In the realm of self-sufficiency and preparedness, the endeavor of raising backyard livestock emerges as a vital aspect. This profound journey encapsulates a harmonious blend of agricultural prowess and practical prepping skills. Delving into the intricate tapestry of sustainable living, this comprehensive guide navigates through the nuances of establishing a thriving backyard livestock operation. As we embark on this exploration, it is imperative to understand the pivotal role that backyard livestock plays in fostering resilience and resourcefulness.

About Backyard Livestock

At the heart of the backyard livestock narrative lies a profound connection between individuals and the animals they raise. It is not merely a venture into animal husbandry; rather, it is a symbiotic relationship that intertwines the livelihoods of both the keeper and the kept. Backyard livestock entails the art of fostering and nurturing a diverse range of creatures, from the humble chicken to the stoic goat and the gentle sheep. Beyond the practicalities of raising animals for sustenance, backyard livestock forms an integral part of a broader lifestyle choice—one that embraces simplicity, sustainability, and a return to the roots of agrarian traditions.

Benefits of Raising Livestock for Preppers

The decision to integrate livestock into a prepper's repertoire is not arbitrary; it is a calculated step towards achieving a myriad of benefits that extend beyond the immediate. One of the foremost advantages lies in the self-sufficiency derived from the production of fresh, homegrown food. The bounty of eggs, milk, and meat sourced from backyard livestock not only reduces reliance on external food sources but also ensures a quality and freshness unparalleled in commercial alternatives.

Moreover, the multifaceted benefits of raising livestock extend to the realm of sustainability. Animals contribute to the ecological balance of a homestead, participating in a cyclical dance of nutrient recycling. Their waste, when managed judiciously, transforms into a potent fertilizer, enhancing soil fertility and promoting the growth of nourishing crops. This interplay of elements fosters a harmonious ecosystem, creating a self-sustaining microcosm within the confines of one's backyard.

The prepper mindset, deeply rooted in foresight and preparedness, finds a natural ally in the cultivation of backyard livestock. By embracing this practice, preppers fortify their strategic approach to survival by diversifying their resource base. In times of unforeseen challenges, the ability to rely on a diversified and resilient source of sustenance becomes paramount. Backyard livestock, with its ability to adapt and thrive, stands as a testament to the foresight embedded in the prepper ethos.

Understanding the Prepper Mindset

To comprehend the prepper mindset is to unravel a tapestry woven with threads of preparedness, resilience, and a commitment to self-reliance. At its core, the prepper mindset is a pragmatic philosophy that acknowledges the unpredictability of the future. It is not rooted in fear but rather in a profound sense of responsibility and empowerment. Preppers are individuals who choose to be architects of their destiny, meticulously crafting contingency plans and building a robust foundation for self-sufficiency.

In the context of backyard livestock, the prepper mindset manifests in a strategic and holistic approach. Every aspect of raising animals is meticulously planned, from the selection of breeds based on climate and resource availability to the construction of robust infrastructure that can withstand unforeseen challenges. The prepper's commitment to learning and adapting is evident in the continuous pursuit of knowledge regarding animal husbandry, sustainable practices, and emergency preparedness.

In conclusion, the integration of backyard livestock into the prepper's arsenal is not just a practical choice; it is a manifestation of a mindset that values preparedness as a cornerstone of a resilient and empowered lifestyle. As we navigate the chapters ahead, the intricate interplay between the essence of backyard livestock and the prepper mindset will continue to unfold, revealing a narrative rich in purpose, sustainability, and the pursuit of a self-reliant existence.

CHAPTER 1

Getting Started with Backyard Livestock

Embarking on the journey of backyard livestock is an exciting venture that demands a meticulous and informed approach. Before delving into the intricacies of animal husbandry, it is imperative to lay a solid foundation by comprehensively assessing the available space and resources. The success of a backyard livestock operation is intricately tied to the suitability of the environment, making space evaluation a crucial starting point.

Assessing Your Space and Resources

The first consideration in getting started with backyard livestock is evaluating the space at your disposal. This involves a holistic assessment of the terrain, climate, and available acreage. Different animals have varied spatial requirements, and understanding these needs is paramount. For example, while chickens may thrive in a smaller backyard setting, goats and sheep typically necessitate more expansive grazing areas. Moreover, considerations such as natural shelter, water sources, and potential hazards must be factored into the space assessment to ensure the safety and well-being of the livestock.

Simultaneously, a thorough examination of existing resources is essential. This includes evaluating the quality of soil for potential grazing or planting, identifying suitable locations for constructing coops or barns, and assessing the availability of water sources. The goal is to create an environment conducive to the health and happiness of the animals while maximizing the use of available resources.

Legal Considerations and Zoning Regulations

Once the spatial evaluation is complete, the next crucial step in getting started with backyard livestock involves navigating the legal landscape. Legal considerations and zoning regulations vary significantly based on geographical location and municipal ordinances. It is imperative to research and understand local laws pertaining to the keeping of livestock.

Zoning regulations often dictate the types and number of animals allowed, as well as the structures permissible on the property. Some areas may have restrictions on certain breeds or species of animals, while others may have noise ordinances that impact the keeping of roosters or other noisy animals. Being well-informed about these regulations helps avoid potential conflicts with local authorities and ensures a smooth and legal establishment of a backyard livestock operation.

Budgeting for Initial Costs

Financial planning is an integral component of getting started with backyard livestock. While the rewards of raising animals can be substantial, there are initial costs associated with setting up infrastructure, acquiring animals, and ensuring their well-being. Budgeting involves a comprehensive breakdown of these costs to facilitate informed decision-making and prevent unexpected financial strain.

Infrastructure costs include the construction of coops, barns, fencing, and any additional structures necessary for the chosen livestock. The quality of materials and construction directly impacts the durability and longevity of these structures. Moreover, acquiring the animals themselves comes with its own set of costs, including purchase or adoption fees, transportation expenses, and initial veterinary care.

Beyond infrastructure and animal costs, ongoing expenses such as feed, bedding, and veterinary care must be factored into the budget. It is prudent to establish a financial cushion for unforeseen circumstances, acknowledging that the initial investment is a long-term commitment. Through meticulous budgeting, prospective backyard livestock keepers can navigate the financial landscape with confidence, ensuring that the journey begins on a stable and sustainable foundation.

With the spatial assessment, legal considerations, and budgeting framework in place, the journey of getting started with backyard livestock extends into the realm of practicality and hands-on preparation. Each facet of this chapter interlaces to provide a comprehensive understanding of the foundational elements necessary for a successful and sustainable livestock venture.

Implementation of Spatial Considerations

The success of a backyard livestock endeavor hinges on the thoughtful implementation of spatial considerations. Once the evaluation is complete, translating insights into practical applications becomes paramount. Designing the layout of the space involves strategic placement of coops, barns, and grazing areas. This spatial design must prioritize ease of access, animal safety, and efficient use of resources.

For smaller spaces, vertical solutions such as multi-tiered chicken coops or vertical grazing structures may be considered. The layout should also account for potential expansion and flexibility to accommodate changes in the livestock setup over time. The implementation phase is an opportunity to optimize the available space, creating an environment where animals can thrive while minimizing the ecological footprint.

Navigating Legal Considerations

Navigating legal considerations demands a proactive and informed approach. This involves communication with local authorities, zoning boards, and other relevant entities. Seeking permits and approvals, when necessary, should be a priority before introducing livestock onto the property. Engaging with the community and neighbors can also foster positive relationships and preemptively address any concerns or misconceptions.

Understanding and adhering to legal requirements is not merely a bureaucratic necessity; it is a fundamental aspect of responsible animal husbandry. Compliance with regulations ensures the well-being of both the animals and the community at large. This process might involve consultations with legal professionals or local agricultural extension offices to gain a comprehensive understanding of the legal landscape.

Fine-Tuning the Budgeting Process

As the spatial and legal aspects take shape, the budgeting process evolves into a dynamic tool for strategic planning. Fine-tuning the budget involves conducting market research to assess the costs of materials, labor, and animal-related expenses. Exploring cost-effective yet durable options for infrastructure and feed without compromising on quality becomes a critical aspect of this phase.

Additionally, contingency planning within the budget is essential. Unforeseen circumstances, such as sudden changes in market prices or unexpected veterinary expenses, can arise. By building flexibility into the budget, potential challenges can be navigated without jeopardizing the stability of the backyard livestock operation.

The budgeting process extends beyond the initial setup costs and encompasses ongoing expenses. An understanding of the recurring costs associated with feed, healthcare, and maintenance ensures that the financial commitment is sustainable in the long run. Through meticulous planning and periodic reviews, the budget becomes a dynamic tool that adapts to the evolving needs of the backyard livestock venture.

Navigating Ahead

In conclusion, Chapter 1 serves as the gateway to the world of backyard livestock, laying the groundwork for a journey infused with practicality and foresight. The meticulous assessment of space and resources, navigation of legal considerations, and strategic budgeting collectively form the pillars on which a successful and sustainable backyard livestock operation stands. As the narrative unfolds, subsequent chapters will delve deeper into the intricacies of selecting and caring for specific livestock breeds, creating robust infrastructure, and embracing the daily rhythms of responsible animal husbandry.

CHAPTER 2

Selecting the Right Livestock for Your Needs

Embarking on the journey of selecting the right livestock for your backyard is a pivotal step that requires a nuanced understanding of the unique characteristics, needs, and benefits associated with different animal species. This chapter dives into the intricate world of three commonly chosen livestock categories: Chickens, Goats, and Sheep. Each species brings its own set of considerations, and a thoughtful selection process is vital for the success and sustainability of a backyard livestock operation.

Chickens: Breeds, Housing, and Care

Chickens, with their diverse breeds and multifaceted contributions, are often the cornerstone of backyard livestock ventures. The first consideration in the realm of chickens involves the selection of suitable breeds. Factors such as egg production, meat quality, temperament, and adaptability to local climates play a crucial role.

Popular choices include Rhode Island Reds for prolific egg-laying, Plymouth Rocks for dual-purpose utility, and Leghorns for efficient egg production.

Once the breeds are chosen, the focus shifts to providing adequate housing and care. Chicken coops should be designed with both comfort and functionality in mind. Features like proper ventilation, nesting boxes, and roosting spaces contribute to the well-being of the flock. Additionally, considerations for predator-proofing the coop are paramount, safeguarding the chickens from potential threats.

Chicken care encompasses a range of practices, including feeding, health monitoring, and egg collection. A balanced diet rich in nutrients, supplemented with calcium for strong eggshells, ensures the overall health of the flock. Regular health checks, vaccinations, and proactive measures against common ailments contribute to the longevity and productivity of the chickens. This comprehensive approach to chicken husbandry sets the stage for a thriving and sustainable poultry component within the backyard livestock framework.

Goats: Types, Feeding, and Shelter

Goats, renowned for their versatility and amiable nature, offer a valuable addition to backyard livestock setups. Understanding the types of goats and their specific characteristics is crucial for selecting breeds that align with your goals. Dairy goats, such as Nubians and Saanens, excel in milk production, while meat breeds like Boers are prized for their robust build and quality meat. Additionally, fiber goats like Angoras contribute to the production of mohair or cashmere.

Feeding goats involves considerations of their dietary needs and foraging habits. While goats are known for their adaptability to different landscapes, providing a balanced diet enriched with roughage, grains, and minerals is essential for optimal health and productivity. Goat shelters should offer protection from the elements, with adequate ventilation to prevent respiratory issues. Elevated platforms for resting and sturdy fencing further contribute to their well-being.

Goat care extends beyond nutrition and shelter to include routine health checks, vaccinations, and parasite control. Understanding the signs of common goat ailments empowers the keeper to take proactive measures, ensuring the longevity and vitality of the herd. By embracing a holistic approach to goat husbandry, individuals can integrate these charming creatures seamlessly into their backyard livestock ensemble.

Sheep: Choosing the Right Breed, Grazing, and Health

Sheep, known for their wool, milk, and meat production, bring a unique dimension to backyard livestock endeavors. Choosing the right breed of sheep is contingent on the specific goals of the keeper. Merinos, prized for their high-quality wool, coexist with meat breeds like Dorsets or Suffolks, known for their robust frames and flavorful meat. Dual-purpose breeds like Shetlands offer a balanced combination of wool and meat.

Grazing management plays a pivotal role in sheep husbandry. Sheep are avid grazers, and rotating pastures prevents overgrazing and promotes optimal forage utilization. The selection and management of grazing areas impact not only the sheep's nutrition but also the overall health of the pasture ecosystem. Strategic grazing practices contribute to sustainable land management, aligning with the principles of responsible animal husbandry.

Sheep health encompasses a range of considerations, from nutrition and vaccination to parasite control. Monitoring the flock for signs of illness, providing necessary vaccinations, and implementing effective parasite prevention measures contribute to the overall well-being of the sheep. Regular shearing, a practice crucial for wool-producing breeds, promotes hygiene and prevents issues such as heat stress.

Navigating Ahead

In conclusion, Chapter 2 illuminates the nuanced process of selecting the right livestock for a backyard venture. Chickens, goats, and sheep each bring their unique contributions to the tapestry of a diversified and sustainable homestead. By understanding the intricacies of breeds, housing requirements, and care practices, individuals can curate a harmonious livestock ensemble that aligns with their goals, values, and the ethos of responsible animal husbandry.

CHAPTER 3

Planning and Building Livestock Infrastructure

The foundation of a successful backyard livestock venture lies in the strategic planning and construction of livestock infrastructure. Chapter 3 delves into the essential elements of this process, examining coop and barn designs, fencing options, as well as watering systems and waste management. Each facet contributes significantly to the overall well-being of the animals, the efficiency of operations, and the sustainability of the homestead.

Coop and Barn Designs

The design and construction of coops and barns are pivotal components of providing a safe and comfortable habitat for the livestock. Coops, designed for smaller animals such as chickens, require thoughtful layouts that balance adequate space, ventilation, and protection from the elements. Nesting boxes, roosting spaces, and easy access for cleaning contribute to the functionality of a well-designed chicken coop.

Barns, on the other hand, cater to larger animals like goats and sheep. The design of a barn encompasses considerations such as proper ventilation, insulation, and sufficient space for animals to move comfortably. Dividing the barn into sections for different purposes—such as feeding, resting, and kidding or lambing—optimizes the use of space and facilitates efficient management.

Both coop and barn designs should prioritize durability and ease of maintenance. The choice of materials, construction techniques, and the inclusion of features like removable panels for cleaning contribute to the longevity of these structures. A well-designed and properly constructed infrastructure not only provides a secure environment for the livestock but also streamlines daily operations for the keeper.

Fencing Options and Considerations

Effective fencing is a cornerstone of responsible animal husbandry, preventing livestock from straying, protecting them from predators, and delineating grazing areas. Chapter 3 explores various fencing options and considerations tailored to the specific needs of different animals. For chickens, lightweight yet sturdy fencing materials such as hardware cloth or welded wire are effective in creating a secure coop environment.

Goats, known for their agility and curiosity, require robust fencing to prevent escapes and protect them from predators. Woven wire or electric fencing, combined with appropriate height and spacing, serves as an effective deterrent. Sheep, while less likely to challenge fences, benefit from similar fencing options, with an emphasis on providing secure enclosures for lambing or shearing activities.

Considerations for fencing extend beyond material selection to encompass factors such as terrain, climate, and local wildlife. Understanding the natural behaviors of the livestock and potential threats from predators informs the design and installation of fences. Regular maintenance and periodic checks ensure the integrity of the fencing, addressing any wear, damage, or potential weak points promptly.

Watering Systems and Waste Management

Ensuring a consistent and clean water supply is paramount to the health and well-being of livestock. Chapter 3 delves into the intricacies of designing effective watering systems that cater to the specific needs of chickens, goats, and sheep. For chickens, nipple or cup waterers placed at a convenient height provide access to clean water while minimizing spillage and contamination.

Goats and sheep benefit from trough or automatic waterers that accommodate their larger size and water consumption. Placement of watering systems in easily accessible locations within grazing areas ensures that animals stay adequately hydrated. Regular cleaning and maintenance of water sources prevent the buildup of contaminants, safeguarding the health of the livestock.

Waste management is an integral aspect of responsible livestock stewardship. Coordinated planning for the collection and disposal of manure minimizes environmental impact and promotes sustainable practices. Composting systems can transform manure into nutrient-rich fertilizer, closing the loop in the waste-to-resource cycle. Adequate drainage in barns and coops, coupled with well-designed manure storage areas, facilitates efficient waste management practices.

Navigating Ahead

In conclusion, Chapter 3 underscores the importance of meticulous planning and construction in creating a robust infrastructure for backyard livestock. Coop and barn designs, fencing options, watering systems, and waste management collectively form the backbone of a sustainable and efficient operation. By carefully considering the unique needs of each species and tailoring infrastructure to those needs, individuals can cultivate a harmonious environment that promotes the health and well-being of the animals while embracing responsible and sustainable practices.

CHAPTER 4

Feeding and Nutrition

The cornerstone of a healthy and thriving livestock operation lies in a deep understanding of feeding and nutrition. Chapter 4 delves into the intricacies of meeting the nutritional needs of various livestock species, emphasizing the importance of creating balanced diets and considering seasonal variations in feeding practices. Nurturing animals with a comprehensive understanding of their dietary requirements ensures not only their individual well-being but also contributes to the overall success and sustainability of the backyard livestock venture.

Understanding Livestock Nutritional Needs

The foundation of effective feeding begins with a thorough comprehension of the nutritional needs specific to each species of livestock. Chickens, goats, and sheep have distinct dietary requirements influenced by factors such as age, breed, reproductive status, and overall health. For chickens, a diet rich in protein is essential for egg production, while goats, being ruminants, require a balanced mix of fiber, carbohydrates, and protein to support their digestive systems.

Sheep, with their diverse roles in wool, milk, and meat production, necessitate nuanced nutritional considerations. High-quality forage and supplements are vital for maintaining optimal body condition and supporting the demands of lactating ewes or growing lambs. Understanding the nutritional content of various feeds and forages, and how they align with the specific needs of each species, is fundamental to creating effective feeding strategies.

Beyond the basics of proteins, carbohydrates, fats, vitamins, and minerals, micronutrients play a crucial role in the overall health of livestock. Trace elements such as selenium, copper, and zinc are essential for metabolic functions and immune system support. A nuanced understanding of these micronutrients allows keepers to fine-tune diets, addressing specific deficiencies and promoting robust health.

Creating Balanced Diets

Achieving optimal nutrition for backyard livestock involves the art of creating well-balanced diets that cater to the unique requirements of each species. Chickens, for instance, thrive on a combination of grains, legumes, green forage, and commercial poultry feed. The inclusion of calcium-rich sources, such as crushed eggshells or oyster shell, supports eggshell formation. Dietary diversity is key to meeting essential nutrient requirements and preventing deficiencies.

Goats, as browsers, benefit from access to a variety of forages, shrubs, and tree leaves. Supplementing with grains and minerals ensures a balanced diet that supports their energy needs and overall health. Sheep, while primarily grazers, also require supplementary feeds during periods of high nutritional demand, such as lactation or late pregnancy. Tailoring diets to meet these specific life stages and roles ensures that each animal receives the nutrients essential for its well-being and productivity.

Incorporating seasonal considerations into diet planning adds another layer of complexity. Foraging opportunities, nutrient content in pastures, and temperature variations influence the dietary needs of livestock throughout the year. During the colder months, providing additional energy-rich feeds helps animals maintain body condition and withstand the challenges of inclement weather. Conversely, in warmer seasons, adjusting diets to accommodate higher forage availability supports natural grazing behaviors.

Seasonal Considerations in Feeding

Seasonal fluctuations impact not only the quantity but also the quality of available forage. Understanding these variations allows keepers to adapt feeding practices accordingly. In winter, when fresh green forage is limited, supplementing with stored hay or silage becomes crucial. The nutritional content of forage can vary based on factors such as soil fertility, plant maturity, and weather conditions, emphasizing the need for periodic forage analysis.

In spring and summer, when pastures are lush with new growth, livestock may have access to a more diverse range of nutrients. However, considerations such as preventing overgrazing, managing parasite challenges, and adjusting mineral supplementation become essential. The ability to navigate these seasonal nuances requires a proactive and flexible approach to feeding management.

Water, often overlooked as a critical component of nutrition, plays a vital role in the digestive processes of livestock. Access to clean and plentiful water is non-negotiable, and keepers must monitor water sources regularly to ensure their quality. Adequate hydration is particularly crucial during hot weather, as animals may consume more water to regulate body temperature.

Navigating Ahead

In conclusion, Chapter 4 sheds light on the intricate world of feeding and nutrition in backyard livestock management. By developing a nuanced understanding of the nutritional needs of chickens, goats, and sheep, keepers can craft balanced diets that support the health, productivity, and resilience of their animals. Incorporating seasonal considerations adds a dynamic dimension to this process, emphasizing the need for adaptability and foresight in meeting the evolving needs of livestock throughout the year. As the narrative unfolds, subsequent chapters will explore health and veterinary care, breeding and reproduction, and the day-to-day aspects of responsible animal husbandry.

CHAPTER 5

Health and Veterinary Care

Ensuring the health and well-being of backyard livestock is paramount for a sustainable and successful operation. Chapter 5 delves into the intricate world of health and veterinary care, exploring the identification of common ailments, vaccination protocols, and the importance of routine health checks. A proactive and informed approach to livestock health not only safeguards the individual animals but also contributes to the overall resilience and longevity of the entire herd or flock.

Identifying Common Ailments

The ability to identify common ailments is a fundamental skill for any livestock keeper. Chickens, goats, and sheep are susceptible to a range of health issues, and early detection plays a pivotal role in successful intervention. For chickens, common ailments include respiratory infections, parasitic infestations, and egg-laying issues. Observing changes in behavior, such as lethargy or reduced egg production, can serve as early indicators of potential health concerns.

Goats, known for their hardy nature, may face challenges such as gastrointestinal parasites, respiratory infections, and hoof issues. Recognizing signs of distress, changes in appetite, or abnormal behavior prompts swift action. Similarly, sheep may experience issues like internal parasites, foot rot, or nutritional deficiencies. Close observation of their demeanor, body condition, and any changes in their wool or coat helps in identifying potential health issues.

Understanding the specific symptoms associated with different ailments empowers keepers to take timely and targeted measures. Regular health checks that involve visual inspections, monitoring vital signs, and noting behavioral changes form a proactive strategy for preventing and addressing common livestock health issues.

Vaccination Protocols

Implementing vaccination protocols is a cornerstone of preventive healthcare in backyard livestock. Vaccines protect animals from contagious diseases, reduce the risk of outbreaks, and contribute to the overall health and productivity of the herd or flock. Chickens benefit from vaccinations against common poultry diseases such as Newcastle disease, infectious bronchitis, and Marek's disease. Tailoring vaccination schedules to the specific needs of the flock and local disease prevalence enhances the efficacy of the immunization program.

Goats and sheep also have specific vaccination requirements based on regional disease risks and individual health considerations. Common vaccinations include protection against clostridial diseases, respiratory infections, and reproductive diseases. Maintaining accurate vaccination records, including dates, types of vaccines administered, and any adverse reactions observed, is crucial for effective disease management and future preventive planning.

The implementation of vaccination protocols goes hand in hand with biosecurity measures. Preventing the introduction of diseases to the premises through quarantine procedures for new animals, limiting access to visitors, and practicing good hygiene minimizes the risk of infectious outbreaks. By combining vaccination strategies with robust biosecurity measures, livestock keepers can fortify their defense against common diseases.

Routine Health Checks

Routine health checks form the foundation of proactive and preventive healthcare. Establishing a regular schedule for visual inspections, monitoring vital signs, and conducting hands-on examinations fosters a deeper understanding of the individual health needs of each animal. For chickens, routine health checks involve assessing feather quality, comb color, and any signs of lethargy or respiratory distress.

Goats and sheep benefit from thorough health checks that include evaluating body condition, assessing hoof health, and checking for signs of internal parasites. Routine examinations also provide an opportunity to monitor reproductive health, ensuring timely interventions for issues such as pregnancy toxemia or dystocia. The inclusion of dental checks and regular deworming further contributes to the overall well-being of these animals.

The holistic approach to routine health checks extends to observing the overall behavior of the livestock. Changes in appetite, social interactions, or patterns of movement may serve as early indicators of underlying health issues. Developing a keen sense of observation and familiarity with the normal behavior of each species enables keepers to detect deviations promptly.

Navigating Ahead

In conclusion, Chapter 5 navigates the intricate terrain of health and veterinary care in backyard livestock management. The identification of common ailments, implementation of vaccination protocols, and the commitment to routine health checks collectively form a robust framework for proactive healthcare. By investing time and attention in understanding the unique health needs of chickens, goats, and sheep, keepers not only safeguard individual animals but also contribute to the resilience and sustainability of their entire livestock ensemble. As the narrative unfolds, subsequent chapters will explore topics such as breeding and reproduction, daily care and maintenance, and maximizing yields from eggs, milk, and meat.

CHAPTER 6

Breeding and Reproduction

The intricate dance of breeding and reproduction lies at the heart of sustainable backyard livestock management. Chapter 6 unravels the complexities of this dynamic process, delving into the understanding of animal reproductive cycles, the art of selective breeding for desirable traits, and the nuances of pregnancy and birthing. Nurturing healthy and productive generations is not only integral to the continuity of a livestock venture but also offers keepers the opportunity to shape and improve the characteristics of their herd or flock.

Understanding Animal Reproductive Cycles

A foundational aspect of successful breeding is a deep comprehension of animal reproductive cycles. Chickens, goats, and sheep exhibit distinct patterns in their reproductive behavior and physiology. Chickens, for instance, follow a regular ovulation cycle, with egg-laying influenced by factors such as age, daylight duration, and environmental conditions. Understanding the natural rhythms of chicken reproductive cycles allows keepers to optimize breeding programs and egg production.

Goats, as seasonal breeders, experience reproductive activity influenced by changes in day length. The onset of estrus, or heat, signals the fertile period when mating is most likely to result in successful pregnancies. A keen observation of goat behavior and physical signs, such as tail wagging and mucous discharge, aids keepers in identifying optimal breeding opportunities.

Sheep, also influenced by seasonal breeding patterns, typically experience increased reproductive activity in the fall. The synchronization of breeding with natural cues ensures that lambs are born during periods of favorable weather and forage availability. Understanding the nuances of sheep reproductive cycles guides keepers in managing breeding schedules and optimizing lambing outcomes.

Selective Breeding for Traits

Selective breeding is an art that allows keepers to shape the genetic traits of their livestock, emphasizing desirable characteristics and traits that align with specific goals. In chickens, selective breeding may focus on enhancing egg-laying productivity, promoting specific feather colors or patterns, or developing breeds tailored for meat production. A meticulous understanding of the genetic traits inherent in different chicken breeds informs the selection process.

Goats, known for their diverse roles in dairy, meat, and fiber production, offer a canvas for selective breeding. Keepers may choose to enhance milk yield, improve meat quality, or develop fiber characteristics by selectively breeding individuals with desirable traits. An understanding of the heritability of traits and the principles of genetic inheritance guides keepers in making informed breeding decisions.

Sheep breeding strategies often involve a combination of traits, including wool quality, meat characteristics, and overall hardiness. Selective breeding allows keepers to develop breeds that thrive in specific climates, exhibit resistance to common diseases, or yield superior wool. The targeted selection of breeding pairs based on their genetic makeup contributes to the gradual improvement of the entire flock over successive generations.

Pregnancy and Birthing

The journey from conception to birthing is a critical phase in the breeding cycle, requiring careful attention and proactive management. Chickens, with their relatively short gestation period, undergo a period of incubation lasting about 21 days. Providing appropriate nesting spaces, ensuring optimal nutrition, and minimizing stress during this period contribute to successful hatching.

Goats, with a gestation period of around 150 days, undergo pregnancy with distinctive stages that demand vigilant care. Monitoring the health of pregnant does, adjusting nutrition to support fetal development, and preparing appropriate kidding spaces are integral aspects of goat pregnancy and birthing management. Recognizing signs of impending labor and providing assistance when needed ensures the well-being of both dam and kids.

Sheep, with a gestation period similar to goats, undergo a carefully managed pregnancy and lambing process. Adequate nutrition, protection from predators, and access to clean and comfortable lambing areas contribute to successful outcomes. Recognizing signs of labor, providing assistance as necessary, and monitoring the health of ewes and lambs post-birth are crucial elements of sheep breeding and reproduction.

Navigating Ahead

In conclusion, Chapter 6 navigates the intricate realm of breeding and reproduction in backyard livestock management. Understanding the nuances of animal reproductive cycles, embracing the art of selective breeding for desirable traits, and managing the intricacies of pregnancy and birthing collectively contribute to the sustainability and improvement of a livestock venture. By cultivating a deep understanding of the unique characteristics and needs of chickens, goats, and sheep, keepers can shape and refine their livestock populations to align with specific goals and enhance the overall resilience of their herds and flocks. As the narrative unfolds, subsequent chapters will explore daily care and maintenance, maximizing yields from eggs, milk, and meat, and the broader considerations of sustainable practices in backyard livestock management.

CHAPTER 7

Daily Care and Maintenance

The essence of successful backyard livestock management lies in the daily care and maintenance bestowed upon the animals. Chapter 7 unravels the intricacies of this routine, exploring the spectrum of daily chores and routine tasks, the importance of grooming and hygiene practices, and the strategies for dealing with emergency situations. These daily rituals not only nurture the health and well-being of chickens, goats, and sheep but also form the backbone of a sustainable and harmonious relationship between keepers and their livestock.

Daily Chores and Routine Tasks

Daily chores form the rhythm of life in a backyard livestock setting, intertwining the needs of the animals with the responsibilities of the keeper. For chickens, the day begins with a visit to the coop for egg collection, ensuring that eggs are promptly gathered to maintain freshness and prevent breakage. Providing fresh water and a balanced feed ensures that the flock starts the day with the essential nutrients for optimal health and productivity.

Goats, known for their inquisitive nature, require daily checks of their living quarters and grazing areas. Routine tasks include ensuring access to clean water, providing a balanced diet that meets their nutritional needs, and inspecting for signs of illness or injury. Maintaining clean and dry bedding in barns or shelters contributes to the overall comfort and well-being of the goats.

Sheep, with their woolly coats and grazing habits, benefit from routine tasks that focus on ensuring access to quality forage, clean water, and suitable shelter. Regular checks for signs of lameness, changes in body condition, or potential health issues allow keepers to address concerns promptly. Daily chores also involve monitoring pregnant ewes for signs of impending labor and providing appropriate care during lambing season.

Grooming and Hygiene Practices

Grooming and hygiene practices are integral components of daily care, contributing to the overall health, comfort, and appearance of backyard livestock. Chickens benefit from dust baths, a natural behavior that aids in controlling mites and lice. Providing areas with fine dust or sand allows chickens to engage in this self-grooming activity, promoting feather health and pest control.

Goats, particularly those with longer coats, may require periodic grooming to prevent matting and tangling. Brushing or combing their coats helps remove loose hair, dirt, and debris. Hoof trimming is another essential grooming task for goats, preventing overgrowth and potential lameness. Additionally, maintaining clean udders in dairy goats contributes to milk quality and udder health.

Sheep, known for their wool production, require meticulous grooming to ensure the quality of their fleece. Regular shearing not only provides valuable wool but also prevents issues such as heat stress and matting. Hoof trimming, similar to goats, is crucial for maintaining proper foot health. Grooming practices also extend to the careful cleaning of udders in ewes, promoting overall hygiene and preventing potential issues during lactation.

Dealing with Emergency Situations

The unpredictability of life with backyard livestock necessitates the development of strategies for dealing with emergency situations. Recognizing signs of illness or distress is a foundational skill for keepers. Changes in behavior, appetite, or vital signs may serve as early indicators of potential issues. Having a basic understanding of common ailments and first aid practices equips keepers to respond swiftly and effectively.

Injuries, such as cuts or wounds, may occur despite vigilant care. Providing immediate attention, cleaning wounds, and applying appropriate treatments help prevent infections and facilitate the healing process. Keepers should maintain a well-stocked first aid kit, including items such as antiseptics, bandages, and medications recommended by a veterinarian.

Pregnancy and birthing in goats and sheep may involve unexpected complications. Keepers should be prepared to assist during labor, recognizing signs of distress and intervening when necessary. Having a designated lambing or kidding area with necessary supplies, such as clean towels, iodine for navel dipping, and a heat lamp for newborns, ensures a smooth transition for both dam and offspring.

Natural disasters, such as storms or fires, may pose challenges to the safety of livestock. Developing contingency plans that include evacuation procedures, access to emergency shelters, and communication strategies with local authorities is paramount. Keepers should also be familiar with evacuation routes, transport options, and the necessary provisions for the well-being of the animals during emergencies.

Navigating Ahead

In conclusion, Chapter 7 unveils the tapestry of daily care and maintenance in backyard livestock management. The intertwining of daily chores and routine tasks, grooming and hygiene practices, and strategies for dealing with emergency situations forms a holistic approach to responsible animal stewardship. By embracing the daily rhythms of life with chickens, goats, and sheep, keepers not only foster the health and happiness of their animals but also cultivate a resilient and harmonious relationship grounded in attentiveness and care. As the narrative unfolds, subsequent chapters will explore the nuances of maximizing yields from eggs, milk, and meat, as well as the broader considerations of sustainable practices in backyard livestock management.

CHAPTER 8

Maximizing Yield: Eggs, Milk, and Meat

Maximizing yields from backyard livestock encompasses a delicate balance between husbandry practices, efficient management, and ethical considerations. Chapter 8 unravels the intricacies of maximizing yield from eggs, milk, and meat, offering insights into egg production tips and tricks, milking techniques, and the processes involved in slaughtering and processing meat. Balancing productivity with the well-being of chickens, goats, and sheep is fundamental to responsible and sustainable practices in backyard livestock management.

Egg Production Tips and Tricks

Egg production is a rewarding aspect of keeping chickens, and optimizing yield requires a blend of strategic management and understanding the nuances of chicken behavior. Providing a well-balanced diet enriched with calcium ensures the hens have the necessary nutrients for robust eggshell formation. Supplementing their diet with oyster shell or crushed eggshells helps prevent calcium deficiencies, reducing the likelihood of soft or brittle eggshells.

Ensuring a comfortable and stress-free environment is essential for egg production. Nesting boxes with clean and soft bedding encourage hens to lay eggs in designated spaces, minimizing the risk of eggs being laid in undesirable locations. Adequate ventilation in the coop, coupled with proper lighting to mimic natural day length, supports the hormonal balance necessary for consistent egg-laying.

The strategic placement of roosting bars and the provision of secure and predator-proof nesting areas contribute to a sense of safety and well-being among the flock, fostering optimal egg production. Collecting eggs regularly prevents broodiness and encourages continued laying. Regular monitoring for signs of illness or stress, coupled with prompt intervention when needed, further supports sustained egg production in backyard flocks.

Milking Techniques and Dairy Management

For those keeping goats, maximizing milk yield involves mastering milking techniques and implementing effective dairy management practices. Milking goats is both a skill and an art, requiring patience, gentle handling, and a keen understanding of goat anatomy. Ensuring a stress-free and comfortable environment during milking sessions contributes to the release of oxytocin, a hormone essential for milk letdown.

Maintaining a consistent milking schedule helps regulate milk production and prevents discomfort for the goats. Adequate nutrition, including access to clean water and a balanced diet rich in nutrients, supports optimal milk yield. The inclusion of a suitable mineral supplement further contributes to the overall health of the lactating does.

Dairy management extends beyond the milking parlor to the proper handling and storage of milk. Prompt cooling and refrigeration prevent bacterial growth and ensure the freshness and quality of the milk. Hygienic practices during milking, such as cleaning udders and teats, reduce the risk of contamination. Implementing routine health checks for lactating does and addressing any signs of illness promptly contribute to sustained milk production.

Slaughtering and Processing Meat

The journey from pasture to plate involves ethical considerations, humane practices, and a meticulous approach to slaughtering and processing meat. For those raising animals for meat production, responsible stewardship includes understanding the principles of humane slaughter and embracing a transparent and respectful approach to the processing phase.

Humane slaughter techniques prioritize the well-being of the animals, minimizing stress and discomfort during the process. This often involves utilizing methods that ensure a swift and painless end, such as stunning followed by a rapid bleed. Adhering to recommended guidelines and regulations for humane slaughter aligns with ethical standards and promotes a compassionate approach to meat production.

Once the animals are slaughtered, the process of meat processing begins. This includes the careful removal of feathers, skin, or wool, depending on the species. Efficient evisceration and cleaning practices are crucial to prevent contamination and ensure the safety of the meat. Following established hygiene and sanitation protocols during the processing phase safeguards the quality and wholesomeness of the meat.

Processing meat also involves butchering and preparing cuts for consumption. A skilled and knowledgeable approach to butchering maximizes the yield of high-quality cuts, minimizing waste. Proper storage and packaging techniques, including freezing or refrigeration, preserve the freshness and flavor of the meat until it reaches the consumer's table.

Navigating Ahead

In conclusion, Chapter 8 unravels the intricate tapestry of maximizing yield from eggs, milk, and meat in backyard livestock management. From implementing egg production tips and tricks for chickens to mastering milking techniques for goats and embracing ethical practices in slaughtering and meat processing, responsible stewardship lies at the core of these endeavors. By balancing productivity with compassion and implementing efficient management practices, keepers can not only maximize the yields from their livestock but also contribute to the sustainability and ethical standards of their backyard operations. As the narrative unfolds, subsequent chapters will explore broader considerations of sustainable practices in backyard livestock management and the interconnected relationship between livestock and the environment.

CHAPTER 9

Utilizing Livestock Products

The cycle of stewardship in backyard livestock management extends beyond the care and maintenance of animals to the thoughtful utilization of their byproducts. Chapter 9 explores the various dimensions of utilizing livestock products, from composting with animal waste to harnessing the nutrient-rich properties of manure as fertilizer. Additionally, creative and sustainable approaches to repurposing other livestock byproducts contribute to a holistic and resourceful model of backyard livestock management.

Composting with Animal Waste

The organic matter generated by chickens, goats, and sheep can be a valuable resource when harnessed through composting. Chicken litter, a combination of manure and bedding materials such as straw or wood shavings, forms an excellent base for compost. The nitrogen-rich content of chicken manure accelerates the composting process, while the carbonaceous bedding materials provide structure and aeration.

In the composting process, microbial activity breaks down the organic matter, transforming it into nutrient-rich humus. Composting with chicken litter yields a potent fertilizer that enhances soil structure, promotes water retention, and enriches the soil with essential nutrients. Incorporating composted chicken litter into garden beds or agricultural fields not only recycles waste but also fosters sustainable soil health.

Similarly, goat and sheep manure, when combined with bedding materials, creates a compost rich in nitrogen, phosphorus, and potassium. This blend contributes to improved soil fertility, enhancing the growth of plants and crops. The composting process not only transforms manure into a valuable resource but also mitigates the potential environmental impacts associated with untreated animal waste.

Using Manure as Fertilizer

Manure, renowned as "black gold" in the realm of agriculture, serves as a potent and natural fertilizer.

Chapter 9 delves into the principles of utilizing chicken, goat, and sheep manure as a sustainable means of enriching soil fertility. The nutrient composition of manure varies by species, offering a diverse array of essential elements for plant growth.

Chicken manure, high in nitrogen, is particularly beneficial for leafy green vegetables and plants with a high demand for this essential nutrient. Applying composted chicken manure or well-aged chicken litter to garden beds or fields provides a slow-release source of nitrogen, promoting healthy plant development and vigorous growth.

Goat and sheep manure, with a balanced blend of nitrogen, phosphorus, and potassium, offers a versatile fertilizer for a variety of crops. The lower nitrogen content compared to chicken manure makes it well-suited for flowering plants, fruit trees, and root vegetables. Incorporating composted goat and sheep manure into the soil enhances nutrient availability, improves soil structure, and supports overall plant health.

Utilizing manure as fertilizer extends beyond its direct application to the soil. Liquid manure, obtained through the process of manure tea or leachate, serves as an effective liquid fertilizer. Diluting the concentrated nutrients in water creates a solution that can be applied directly to plants, providing an additional avenue for harnessing the benefits of livestock manure.

Other Creative Ways to Utilize Livestock Byproducts

Beyond composting and using manure as fertilizer, backyard livestock byproducts offer a plethora of creative and sustainable applications. The utilization of feathers, wool, and milk byproducts showcases the resourcefulness inherent in responsible livestock stewardship.

Feathers shed by chickens during molting or shearing sheep wool can be repurposed for various purposes. Feathers are rich in nitrogen, making them suitable for composting. When properly composted, feathers contribute to nutrient-rich humus that enhances soil fertility. Additionally, feathers can be incorporated into the production of nutrient-rich teas or used as mulch to suppress weeds and retain soil moisture.

Wool, a valuable byproduct of sheep, possesses unique insulating properties. Utilizing sheared wool for composting or mulching aids in water retention, weed suppression, and temperature regulation in the soil. Wool fibers can also be repurposed for crafts or used as a natural mulch for potted plants.

Milk byproducts, such as whey, offer diverse applications beyond consumption. Whey, a byproduct of cheese-making, serves as an excellent liquid fertilizer. Its nutrient-rich composition, including proteins, lactose, and minerals, contributes to plant growth and soil health. Diluting whey with water creates a nutrient solution that can be applied to both garden and field crops.

Navigating Ahead

In conclusion, Chapter 9 highlights the resourceful and sustainable utilization of livestock products in backyard livestock management. From composting with animal waste to using manure as fertilizer and exploring creative applications for feathers, wool, and milk byproducts, responsible stewardship extends beyond the animals themselves. By harnessing the valuable byproducts generated by chickens, goats, and sheep, keepers not only contribute to the fertility and health of their soil but also engage in a model of holistic and sustainable backyard livestock management. As the narrative unfolds, subsequent chapters will explore broader considerations of sustainable practices in backyard livestock management and the interconnected relationship between livestock and the environment.

Chapter 10

Sustainable Practices and Future Considerations

Sustainable practices form the cornerstone of responsible backyard livestock management, transcending immediate care to encompass a thoughtful and eco-friendly approach. Chapter 10 navigates the landscape of sustainable practices and future considerations, guiding keepers on implementing environmentally conscious methods, fostering long-term planning for sustainable livestock endeavors, and exploring opportunities for expanding or diversifying their operations.

Implementing Eco-Friendly Practices

Sustainable practices in backyard livestock management go hand in hand with environmental stewardship. From energy conservation to waste reduction, implementing eco-friendly practices contributes to the overall health of the ecosystem in which chickens, goats, and sheep coexist. The chapter unravels various dimensions of sustainable practices, shedding light on areas such as energy efficiency, waste management, and habitat conservation.

Efficient energy use begins with the design and maintenance of livestock infrastructure. Implementing energy-efficient lighting, utilizing renewable energy sources like solar power, and employing insulation measures contribute to minimizing the ecological footprint of backyard livestock operations. Beyond infrastructure, incorporating natural lighting in coop and barn designs aligns with the natural rhythms of animal behavior, promoting overall well-being and minimizing the need for artificial lighting.

Waste management is a critical aspect of sustainability, and responsible disposal of animal waste plays a pivotal role. Composting, as explored in previous chapters, remains a sustainable approach to repurposing manure and bedding materials. Additionally, exploring vermiculture or worm composting systems leverages the natural processes of earthworms to break down organic matter, creating nutrient-rich vermicompost that further enriches soil.

Habitat conservation within the backyard setting enhances biodiversity and ecological balance. Creating designated areas with native plants and natural vegetation supports local pollinators, beneficial insects, and small wildlife. This intentional approach to landscaping contributes to a harmonious coexistence between livestock and the broader ecosystem.

Water conservation practices form another dimension of sustainability. Implementing efficient watering systems, capturing and repurposing rainwater, and promoting responsible water usage contribute to overall resource conservation. Adequate hydration for chickens, goats, and sheep is maintained while minimizing water wastage and environmental impact.

Long-Term Planning for Sustainable Livestock

Sustainable livestock management extends beyond immediate practices to encompass thoughtful long-term planning. Chapter 10 delves into the strategies for fostering sustainability over the years, ensuring the well-being of animals, the resilience of the operation, and the preservation of natural resources.

Breeding strategies play a vital role in long-term sustainability. Selective breeding for traits such as disease resistance, climate adaptability, and overall hardiness contributes to the development of robust livestock populations. Understanding the genetic makeup of individual animals and making informed decisions about breeding pairs aligns with the goals of maintaining healthy and resilient herds and flocks.

The integration of permaculture principles within the livestock environment fosters a holistic and sustainable approach to land use. Designing landscapes that mimic natural ecosystems, incorporating diverse plant species, and utilizing rotational grazing practices optimize soil health and forage availability. This approach promotes the symbiotic relationship between livestock and the land, allowing for regenerative practices that enhance biodiversity and reduce the impact of overgrazing.

Implementing biosecurity measures ensures the long-term health and well-being of livestock. Preventing the introduction of diseases through quarantine procedures, limiting contact with outside animals, and maintaining good hygiene practices minimize the risk of infectious outbreaks. Biosecurity, coupled with routine health checks and vaccination protocols, contributes to the overall resilience of the livestock population.

Long-term planning also involves anticipating the changing needs of the livestock operation. Considering factors such as aging animals, evolving market demands, or shifts in personal circumstances enables keepers to adapt and make informed decisions about the future of their livestock endeavors. Developing contingency plans for emergencies or unexpected challenges enhances the overall sustainability of the operation.

Expanding or Diversifying Your Livestock Endeavors

The journey of backyard livestock management is dynamic, offering opportunities for expansion or diversification. Chapter 10 explores the considerations and strategies for those contemplating the growth or diversification of their livestock endeavors. Whether motivated by increased interest, evolving goals, or market opportunities, expanding or diversifying requires thoughtful planning and a commitment to sustainable practices.

Expanding the livestock operation may involve increasing the size of the existing herd or flock. Considerations include assessing available space, ensuring adequate resources for additional animals, and evaluating the capacity of existing infrastructure. Thoughtful planning, including considerations for breeding strategies, nutritional requirements, and health management, ensures a seamless integration of new members into the livestock ensemble.

Diversification opens the door to exploring different species of livestock or incorporating additional facets such as agroforestry, beekeeping, or aquaculture. Introducing new species should be guided by a thorough understanding of the unique needs and behaviors of each, as well as their compatibility with existing livestock. Thoughtful integration promotes a harmonious and sustainable multi-species environment.

Diversifying product offerings, such as incorporating value-added products like artisanal cheeses, wool crafts, or specialty eggs, adds layers of complexity and opportunity to a livestock operation. Market research, understanding consumer demand, and developing skills or partnerships necessary for product diversification contribute to the overall resilience and sustainability of the enterprise.

Navigating Ahead

In conclusion, Chapter 10 weaves a narrative of sustainable practices and future considerations in backyard livestock management. From implementing eco-friendly practices and fostering long-term planning to exploring opportunities for expanding or diversifying operations, responsible stewardship extends into the realm of thoughtful and intentional decision-making. By embracing sustainability as a guiding principle, keepers not only nurture the health and well-being of their animals but also contribute to the resilience and harmony of the broader ecosystem. As the narrative unfolds, the holistic tapestry of backyard livestock management continues to evolve, embracing the interconnected relationship between livestock and the environment.

CONCLUSION

Embarking on the journey of backyard livestock management is a venture that intertwines the practicalities of animal care with the broader principles of sustainability and responsible stewardship. As we draw the final threads of this comprehensive guide, the overarching tapestry reveals a rich narrative of insights, strategies, and considerations that form the foundation of successful and fulfilling livestock endeavors. From the initial stages of preparation and choosing the right livestock to the intricate details of breeding, nutrition, and health care, each chapter has woven a fabric of knowledge for aspiring and seasoned keepers alike.

Recap of Essential Points

In reflection, the guide began by delving into the nuances of preparation and the critical considerations that precede the introduction of livestock to the backyard setting. Assessing space, navigating legal considerations, and budgeting for initial costs emerged as foundational steps, ensuring that keepers embark on their journey with clarity and foresight. Subsequent chapters navigated the intricacies of selecting the right livestock, planning and building infrastructure, and addressing the daily care and maintenance essential for the well-being of chickens, goats, and sheep.

The exploration of breeding and reproduction unfolded as a pivotal chapter, unraveling the mysteries of animal reproductive cycles, selective breeding for traits, and the delicate art of pregnancy and birthing. As the guide transitioned to maximizing yields from eggs, milk, and meat, readers gained insights into the finer details of egg production, milking techniques, and the responsible practices involved in slaughtering and processing meat.

The utilization of livestock products revealed the resourceful dimensions of backyard management, from composting with animal waste to using manure as fertilizer and exploring creative applications for feathers, wool, and milk byproducts. As keepers embraced the potential of these byproducts, a sustainable model of livestock management emerged, highlighting the interconnected relationship between animals and the environment.

The final chapters steered the narrative toward sustainability and future considerations. Implementing eco-friendly practices, fostering long-term planning for sustainable livestock, and contemplating opportunities for expanding or diversifying operations underscored the holistic and forward-thinking approach required for lasting success. From energy efficiency and waste reduction to biosecurity measures and diversified product offerings, the guide illuminated the path toward a resilient and harmonious coexistence between livestock and their keepers.

Encouragement for Ongoing Success in Backyard Livestock Management

As keepers embark on or continue their journey in backyard livestock management, the guide concludes with a resonant note of encouragement. Success in this endeavor is not only measured by the productivity of the flock or herd but by the harmonious relationship cultivated between keeper and animals. The daily rhythms of care, the thoughtful utilization of byproducts, and the integration of sustainable practices contribute not only to the well-being of the animals but also to the health of the land and the broader environment.

Ongoing success in backyard livestock management rests on the foundation of knowledge, empathy, and adaptability. Keepers are encouraged to embrace a continuous learning mindset, staying abreast of advancements in animal husbandry, veterinary care, and sustainable practices. Networking with fellow keepers, participating in community events, and seeking guidance from agricultural extension services fosters a supportive and collaborative environment where knowledge is shared and experiences are exchanged.

Adaptability is a key attribute in the journey of livestock management. As seasons change, animals age, and external factors come into play, the ability to adapt management practices ensures resilience in the face of challenges. The principles of sustainability and responsible stewardship serve as guiding stars, providing a compass for decision-making that aligns with the long-term health of the animals and the environment.

The bond between keeper and livestock is a reciprocal relationship, and the animals become not just a part of the backyard but integral members of a shared ecosystem. Observing their behavior, responding to their needs, and fostering an environment that allows them to express natural behaviors enhances the overall quality of life for both keeper and animals. This shared journey is a testament to the profound connection that can be forged between humans and the creatures they steward.

In conclusion, as the threads of knowledge and experience weave together, keepers are poised to create a vibrant and sustainable tapestry of backyard livestock management. The chapters explored in this guide serve as stepping stones, offering insights, strategies, and considerations to inform and inspire keepers on their journey. The landscape of backyard livestock management is dynamic, evolving with each season and challenge. May this guide stand as a companion, providing guidance, encouragement, and a source of inspiration as keepers continue to nurture and thrive in their roles as stewards of backyard livestock.